Introduction to THE THEORY OF DISTRIBUTIONS

Canadian Mathematical Congress, Lecture Series, No. 1

INTRODUCTION TO THE THEORY OF DISTRIBUTIONS

by

ISRAEL HALPERIN
Associate Professor of Mathematics, Queen's University

based on the lectures given by

LAURENT SCHWARTZ
Professor of Mathematics, University of Nancy

UNIVERSITY OF TORONTO PRESS
TORONTO

Printed in Canada
London: Oxford University Press
Reprinted, 1960, 1968, 2017
ISBN 978-1-4875-9132-8 (paper)

PREFACE

The Theory of Distributions was the subject of a course of lectures given in the Seminar of the Canadian Mathematical Congress held in Vancouver, August-September 1949. Since then my books have appeared and it does not seem useful to give a summary reproducing exactly my Canadian lectures. Instead, this pamphlet gives a detailed introduction, in terms of classical analysis, for applied mathematicians and physicists. Further study, with my books, requires some knowledge of functional-theoretic analysis. This explains the length of the development of the basic ideas and the brief mention of convolution and Fourier series.

The pamphlet was written by Professor Halperin whom I thank very much. He was led to think through the main problems again and many conceptions here are more Professor Halperin's than mine.

LAURENT SCHWARTZ

CONTENTS

THE THEORY OF DISTRIBUTIONS

§1. Introduction

Since the introduction of the operational calculus at the end of the last century, many formulae have been used which have not been adequately clarified from the mathematical point of view. For instance, consider the Heaviside function $Y(x)$ which vanishes for values of x not exceeding zero and is equal to 1 for positive x. It is said that the derivative of this function is the Dirac delta-function $\delta(x)$ which has the following (mathematically impossible!) properties: it vanishes everywhere except at the origin where its value is so large that

$$\int_{-\infty}^{+\infty} \delta(x)dx = 1.$$

This "function" and its successive "derivatives" have been used with considerable success.

It has been suggested by Dirac himself that the delta-function could be avoided by using instead a limiting procedure involving ordinary (mathematically possible) functions. However, the delta-function can be kept and made rigorous by defining it as a measure, that is, as a set-function in place of an ordinary point-function. This suggests that the notion of point-function be enlarged to include new entities[1] and that the notion of derivative be correspondingly generalized so that within the new system of entities, every point-function should have a rigorously defined derivative. This is done in the theory of distributions. The new system of entities, which we call distributions, includes all continuous functions, all Lebesgue locally summable functions and new objects of which a simple example is the Dirac measure function mentioned above. The more general (but rigorous) process of derivation assigns to every distribution a derivative which is again a distribution and so every distribution, including every locally summable point-function, has derivatives of all orders. The derivative of a locally summable point-function is always a distribution although not, in general, a point-function. However, it coincides with the classical derivative when the latter exists and is locally summable.

This theory of distributions gives rigorous content and validity to the formulae of operational calculus mentioned above. It can be developed not only for functions of one variable but also for functions of several variables and it provides a simple but more complete theory of such topics as Fourier Series and Integrals, Convolutions, and Partial Differential Equations.

[1]Just as the notion of rational number was enlarged by Dedekind to include all real numbers.

A systematic exposition of the theory of distributions is given in *Théorie des Distributions* by Laurent Schwartz, published by Hermann et Cie, Paris as Nos. 1091 (Tome I, 1950) and 1122 (Tome II, 1951) of the series, Actualités Scientifiques et Industrielles. The following pages may however serve as a useful introduction to some of the basic ideas.

§2. Point-functions as functionals

THE following discussion will lead to the precise definition of distributions given later in this section.

Let (a, b) be a finite closed interval. A continuous $f(x)$ can be considered as a point-function[2] but it can also be considered in another way:[3] it defines a *functional* $F(\phi)$ by the formula

$$F(\phi) = \int_a^b f(x)\phi(x)dx$$

where $F(\phi)$ is a number[4] defined for every continuous $\phi(x)$. And this functional is linear, that is

$$F(c_1\phi_1 + c_2\phi_2) = c_1F(\phi_1) + c_2F(\phi_2).$$

Since there are linear functionals which cannot be expressed in this way in terms of any continuous (or even merely Lebesgue summable $f(x)$), our first suggestion is that the distributions be defined as the (arbitrary) linear functionals $F(\phi)$ over some suitable set S of continuous $\phi(x)$. The $\phi(x)$ chosen will be called testing functions.

If $f(x)$ is absolutely continuous and has a derivative $f'(x)$, the derivative will also define a linear functional, namely,

$$\int_a^b f'(x)\phi(x)dx.$$

This new functional can be expressed, for some ϕ, in terms of the original $F(\phi)$ by use of integration by parts. The formula

$$\int_a^b f'(x)\phi(x)dx = -\int_a^b f(x)\phi'(x)dx = -F(\phi')$$

is valid for all $\phi(x)$ which vanish at a and at b and have a continuous first derivative. This suggests: first, that we allow S to include only such $\phi(x)$

[2]Point-functions need be defined only almost everywhere and two functions are identified if they are equal almost everywhere. Thus we say that $f(x)$ is identically zero if it has the value zero almost everywhere, that $f(x)$ is continuous if it can be identified with a continuous function; the upper bound of $f(x)$ will mean the essential upper bound, the derivative of $f(x)$ will mean the derivative of that function which can be identified with $f(x)$ and which has a derivative if such a function exists, and so on.

[3]Just as every rational number was considered by Dedekind in a new way: as defining a cut in the set of all rationals.

[4]Numbers may be taken to mean real numbers or complex numbers.

and not all continuous $\phi(x)$; second, that for every linear $F(\phi)$, whether it comes from an $f(x)$ possessing a derivative $f'(x)$ or not, a derivative of F, to be written as $F'(\phi)$ (a linear functional on S), be defined by

$$F'(\phi) = - F(\phi').$$

But if $F'(\phi)$ is to be defined for all ϕ in S we will have to restrict S to include only such $\phi(x)$ as have $\phi'(x)$ also in S. This leads to the condition: $\phi(x)$ shall have derivatives of all orders and they, as well as $\phi(x)$ itself, shall vanish at a and at b.

Important examples of such $\phi(x)$ are the functions

$$\{\phi_{c,d}(x)\}^{1/n}$$

obtained by choosing any positive integral n and $a \leqslant c < d \leqslant b$ and defining

$$\begin{aligned} \phi_{c,d}(x) &= e^{-\left(\frac{1}{x-c}+\frac{1}{d-x}\right)} && \text{for } c < x < d, \\ &= 0 && \text{for all other } x. \end{aligned}$$

We shall not find it necessary to restrict S further. However, in restricting S we must be careful of one point: we intend to "identify" the point-function $f(x)$ and the functional $F(\phi)$ which it defines. Thus we want S to include sufficiently many testing functions so that functions $f(x)$ which are different on (a, b) will be identified with different $F(\phi)$. What is the same thing, we do not want

$$\int_a^b f(x)\phi(x)dx$$

to vanish for all ϕ in S except when $f(x)$ is identically zero on (a, b). Now this requirement is satisfied if S includes all

$$\{\phi_{c,d}(x)\}^{1/n},$$

described above, for

$$\lim_{n\to\infty} \int_a^b f(x)\{\phi_{c,d}(x)\}^{1/n}dx = \int_c^d f(x)dx;$$

and if

$$\int_c^d f(x)dx = 0$$

for all $a \leqslant c < d \leqslant b$ then $f(x)$ is identically zero on (a, b).

Another point is this: the $F(\phi)$ which are defined by point-functions, and all derivatives of such (the $F^{(n)}(\phi)$) have a certain continuity property and it is desirable to require this continuity property in defining distributions.[5]

[5]As we shall show in §5, this gives the smallest enlargement or completion of the system of locally summable point-functions which permits unrestricted derivation.

Finally, with a view to later applications, we shall define distributions on open intervals. The closed interval gives the mathematically simpler situation and indeed the open interval will be discussed in terms of its closed subintervals. However, we shall reserve the word "distribution" for the open interval and use the terminology "continuous linear functional" for the closed interval.

Our precise definitions now follow.

Definition of distribution. For any closed finite interval (a, b) let $S_{(a,b)}$ consist of all continuous $\phi(x)$ possessing derivatives $\phi^{(n)}(x)$ of all orders which, along with $\phi(x)$ itself, vanish at a and at b and for x outside (a, b). A linear functional $F(\phi)$, defined for all ϕ in $S_{(a,b)}$, is called a c.l.f. on (a, b) if it has the following continuity property; whenever all ϕ, ϕ_m are in $S_{(a,b)}$ and the $\phi_m(x)$ converge uniformly to $\phi(x)$, and, for each n, the derivatives $\phi_m^{(n)}(x)$ converge uniformly to $\phi^{(n)}(x)$, then $F(\phi_m)$ shall converge to $F(\phi)$.

For an arbitrary open interval I, finite or infinite, a distribution on I is a linear functional $F(\phi)$ such that for every closed finite interval (a, b) contained in I, $F(\phi)$ is defined for all ϕ in $S_{(a,b)}$ and, when restricted to these ϕ, defines a c.l.f. on (a, b).

Identification of distribution and point-function. A distribution $F(\phi)$ on an interval I is to be identified with a point-function $f(x)$ if for every closed finite interval (a, b) contained in I, $f(x)$ is summable on (a, b) and

$$F(\phi) = \int_a^b f(x)\phi(x)dx$$

for all ϕ in $S_{(a,b)}$. We shall sometimes use the notation $f(\phi)$ to denote the distribution identified with the point-function $f(x)$.

Identification of distribution and measure function. A distribution $F(\phi)$ on an interval I is to be identified with the Stieltjes measure $d\psi(x)$ if for every closed finite interval (a, b) contained in I, $\psi(x)$ is of bounded variation on (a, b) and

$$F(\phi) = \int_a^b \phi(x)d\psi(x)$$

for all ϕ in $S_{(a,b)}$.

Definition of derivative of a distribution. For any distribution $F(\phi)$ on I, the derivative-distribution $F'(\phi)$ is defined by

$$F'(\phi) = -F(\phi').$$

It is easily verified that this F' satisfies the conditions for a distribution on I. (The same formula defines derivative for a c.l.f.)

The distributions defined above include all continuous and even all Lebesgue (locally) summable point-functions, all Stieltjes measures, and, as we shall see, a variety of new mathematical entities. Within the system of distributions each distribution has a derivative and consequently, derivatives of all orders:

$$F^{(n)}(\phi) = (-1)^n F(\phi^{(n)}).$$

The derivative of a point-function $f(x)$ may be a point-function or a Stieltjes measure or a more general distribution. It will be a point-function $g(x)$ if and only if $f(x)$ is absolutely continuous on every finite closed (a, b) contained in I and then $g(x)$ is the ordinary point-derivative $f'(x)$ which then must be defined (almost everywhere). The derivative of $f(x)$ is a Stieltjes measure $d\psi(x)$ if and only if $f(x)$ is of bounded variation on every finite closed (a, b) contained in I and then $\psi(x)$ differs from $f(x)$ by an additive (arbitrary) constant.

In particular, the derivative δ of the Heaviside function $Y(x)$ is now rigorously defined (I being any open interval which contains the origin) and is the measure function $dY(x)$; this is the proper mathematical description of the Dirac delta-function which has no meaning as a point-function. We may now use the notation: $\delta = Y'$.

In applications to certain physical problems a locally summable $f(x)$ may be thought of as representing a distribution of mass or electric charge on the axis of x:

$$\int_c^d f(x)dx$$

is then the total (algebraic) charge on (c, d) and $f(x)$ is the density of charge at a particular x. From this point of view the Dirac δ represents the concentration of unit charge at a single point, the origin, δ' represents a dipole and higher derivatives of δ represent more complicated multiple-layers.

An interesting distribution which is quite different from the Dirac δ and its derivatives is this: let $f(x)$ have the value $x^{-\frac{1}{2}}$ for positive x and vanish for all other x. Then its derivative f' exists (as a distribution) although it is not quite a Stieltjes measure. Roughly speaking, f' corresponds to negative mass continuously distributed on the positive x axis with an infinite quantity in every neighbourhood of the origin, together with an infinite positive mass at the origin, in such a way that there is finite total algebraic mass on every finite closed interval. This is shown by the formula, for ϕ in $S_{(a,b)}$ with $a < 0 < b$:

$$f'(\phi) = -\int_a^b f(x)\phi'(x)dx = -\int_0^b f(x)\phi'(x)dx$$

$$= -\lim_{\epsilon\to 0}\int_\epsilon^b x^{-\frac{1}{2}}\phi'(x)dx$$

$$= -\lim_{\epsilon\to 0}\left[(x^{-\frac{1}{2}}\phi(x))_\epsilon^b + \int_\epsilon^b \tfrac{1}{2}x^{-3/2}\phi(x)dx\right]$$

$$= \lim_{\epsilon\to 0}\left[\frac{\phi(\epsilon)}{\sqrt{\epsilon}} + \int_\epsilon^b (-\tfrac{1}{2}x^{-3/2})\phi(x)dx\right]$$

$$= \lim_{\epsilon\to 0}\left[\frac{\phi(0)}{\sqrt{\epsilon}} + \int_\epsilon^b (-\tfrac{1}{2}x^{-3/2})\phi(x)dx\right]$$

since

$$\frac{\phi(\epsilon)-\phi(0)}{\sqrt{\epsilon}} = \sqrt{\epsilon}\,\frac{\phi(\epsilon)-\phi(0)}{\epsilon} \to 0.\phi'(0) = 0 \text{ as } \epsilon\to 0.$$

Although $-\frac{1}{2}x^{-3/2}\phi(x)$ is not summable on $(0, b)$ for arbitrary ϕ in $S_{(a,b)}$, yet the bracket as a whole always has a finite limit, which has been called by Hadamard the "finite part" of the divergent integral. We use the notation

$$f'(\phi) = \text{Fp.}\int_0^b f'(x)\phi(x)dx$$

where $f'(x)$ is the non-summable point-derivative of $f(x)$ for positive x. Finite parts of divergent integrals have been studied in great detail by Hadamard.[6]

Similarly, the linear functional

$$F(\phi) = \lim_{\epsilon\to 0}\left[\left(\int_a^{-\epsilon} + \int_\epsilon^b\right)\frac{1}{x}\phi(x)dx\right]$$

corresponds to a continuously distributed mass with infinite positive mass along the positive x axis together with infinite negative mass along the negative x axis in such a way that in every neighbourhood of the origin there is finite total algebraic mass. The limit of the bracket is another case of an Hadamard finite part of a divergent integral, in this case coinciding with the Cauchy principal value. The distribution itself is the derivative of the point-function $\log|x|$.

[6] J. Hadamard, *Le problème de Cauchy et les équations aux dérivées partielles linéaires hyperboliques*, Paris, Hermann et Cie, 1932.

§3. THE CALCULUS OF DISTRIBUTIONS

MULTIPLICATION of a distribution by a constant and addition of two distributions are defined by the formulae:

$$(cF)(\phi) = cF(\phi), \qquad (F_1 + F_2)(\phi) = F_1(\phi) + F_2(\phi).$$

It is easily verified that the usual rules of addition and subtraction hold, even when derivation is involved. Thus

$$(c_1F_1 + c_2F_2)' = c_1F_1' + c_2F_2'.$$

The point-function which is a constant, $f(x) = k$, is identified with constant-distribution $k(\phi)$ which has the characteristic property:

$$k(\phi) = \int_a^b k\phi(t)dt = k\int_a^b \phi(t)dt$$

for every ϕ in $S_{(a,b)}$. In particular, when $k = 0$ the corresponding distribution is called the zero-distribution. It is easily verified that if F is a constant-distribution F' is the zero-distribution. To prove the converse, it is useful to establish the following *expansion* lemma.

Lemma. *Let $\theta(x)$ be any function in $S_{(c,d)}$ for which*

$$\int_c^d \theta(x)dx = 1$$

(for instance $\theta(x)$ could be

$$\left\{\int_c^d \phi_{c,d}(t)dt\right\}^{-1} \phi_{c,d}(x)$$

where $\phi_{c,d}(x)$ is the function defined in §2). Let n be any positive integer. Then any $\phi(x)$ in any $S_{(a,b)}$ with $a \leqslant c < d \leqslant b$ can be expressed in the form:

$$\phi(x) = a_0\theta(x) + a_1\theta'(x) + \ldots + a_n\theta^{(n)}(x) + \rho_n^{(n+1)}(x)$$

where $a_0, a_1, \ldots, a_n$ are uniquely determined constants and $\rho_n(x)$ is in $S_{(a,b)}$.

This lemma can be proved by induction on n if we observe that a particular $\phi(x)$ in $S_{(a,b)}$ can be expressed in the form $\rho'(x)$ for some $\rho(x)$ in $S_{(a,b)}$ if and only if

$$\int_a^b \phi(t)dt = 0.$$

In the expansion for a general $\phi(x)$ in $S_{(a,b)}$,

$$a_0 = \int_a^b \phi(t)dt.$$

Now using this representation with $n = 1$, we argue that if $F' = 0$ then $F(\rho_1') = -F'(\rho_1) = 0$ for all ρ_1, and hence

$$F(\phi) = F(a_0\theta + \rho_1') = a_0F(\theta) + F(\rho_1') = a_0k$$

where k is a constant, the value of $F(\theta)$. Hence

$$F(\phi) = \int_a^b k\phi(t)dt,$$

proving that F is indeed a constant-distribution.

We shall say that the distribution G is a *primitive* of F if $G' = F$. From the preceding paragraph it follows that two primitives of the same F differ by a constant-distribution.

As is well known, in the case of point-functions there exists an infinity of different (point-function) primitives and in order to specify one of them it is sufficient to give its value at some particular point. In the case of an arbitrary distribution F there exists again an infinity of primitives (distributions!) and a particular one can be specified by giving its value for any particular testing-function $\theta(x)$ as described above; this follows immediately from the relationship:

$$G(\phi) = G(a_0\theta + \rho_1') = a_0G(\theta) + G(\rho_1') = a_0G(\theta) - F(\rho_1)$$

which can be used to define $G(\phi)$ for every ϕ in terms of an arbitrary (but fixed) $G(\theta)$, and given F; that this G is a distribution can be deduced from the relation

$$\rho_0(x) = \int_a^x \phi(t)dt - \left\{\int_a^b \phi(t)dt\right\}\int_a^x \theta(t)dt.$$

§4. Multiplication of distributions

The product F_1F_2 is *not* defined for arbitrary distributions. This reflects the fact that the product $f_1(x)f_2(x)$ of two locally summable point-functions may not be locally summable.

However, we do define F_1F_2 in certain cases. For example, if F_1, F_2 can be identified with $f_1(x)$, $f_2(x)$ respectively and the product $f_1(x)f_2(x)$ is locally summable then F_1F_2 is defined to be the distribution which is identified with $f_1(x)f_2(x)$. This is a special case of the following general definition.

Definition of product of two c.l.f.'s on a finite closed interval. Suppose for some n that $F_1^{(n)}$ is a point-function $f_1(x)$ and that F_2 is the nth derivative of a point-function $f_2(x)$:

$$F_1^{(n)} = f_1, \quad F_2 = f_2^{(n)}.$$

Suppose, too, that the product $f_1(x)f_2(x)$ is summable. Then we define F_1F_2 by the formula:

$$F_1F_2 = F_1f_2^{(n)} = (F_1f_2)^{(n)} - \binom{n}{1}(F_1'f_2)^{(n-1)} + \binom{n}{2}(F_1^{(2)}f_2)^{(n-2)}$$
$$+ \ldots + (-1)^r\binom{n}{r}(F_1^{(r)}f_2)^{(n-r)} + \ldots + (-1)^nF_1^{(n)}f_2.$$

Each term on the right is a c.l.f. since $F_1^{(n)}f_2$ is $f_1(x)f_2(x)$ which was assumed summable and for $r < n$, $F_1^{(r)}f_2$ is the product of an absolutely continuous (hence bounded) point-function and f_2.

Definition of product of distributions on an open interval I. Let $F_{(a,b)}$ denote the distribution F restricted to the testing functions in $S_{(a,b)}$. If now F_1, F_2 are distributions on I such that $F_{1(a,b)}F_{2(a,b)}$ is defined on every (a, b) contained in I, then F_1F_2 is defined to be the distribution on I which, when restricted to any $S_{(a,b)}$, coincides with $F_{1(a,b)}F_{2(a,b)}$.

It is not difficult to verify that these definitions give a unique F_1F_2 whenever they define F_1F_2 at all and that our formula above for the product F_1F_2 is equivalent to:

$$(F_1F_2)(\phi) = (-1)^n \int_a^b f_2(x)(F_1(x)\phi(x))^{(n)}dx.$$

It can also be verified that the product law $(F_1F_2)' = F_1'F_2 + F_1F_2'$ is valid and that the three products which occur in this statement are necessarily defined whenever one of the products on the right is defined.

There is a special but most important case in which our product rule can be greatly simplified. Suppose that F_1 is a continuous point-function $\alpha(x)$ possessing point-function derivatives of all orders so that $F_1^{(n)}$ is a continuous point-function for each n. Then our product rule defines F_1F_2 for every F_2 which, when restricted to an $S_{(a,b)}$, can be put in the form $f_2^{(n)}$ for some f_2 and n which might depend on a and b. Our rule simplifies in this case to:

$$(F_1F_2)(\phi) = (\alpha F_2)(\phi) = F_2(\alpha\phi)$$

for every testing function ϕ. (Observe that $\alpha\phi$ is a testing function along with ϕ.)

The preceding paragraph suggests that we could use the relation

$$(\alpha F_2)(\phi) = F_2(\alpha\phi)$$

to *define* αF_2 whenever $\alpha(x)$ has (ordinary) derivatives of all orders but F_2 is an *arbitrary* distribution. It is noteworthy that such a definition would not give anything new since *every* distribution F_2 when restricted to an $S_{(a,b)}$, can be put in the form $f_2^{(n)}$ for some point-function $f_2(x)$ and some n, which may depend on a and b. This theorem will be proved in the next section and it is a consequence of the continuity condition in our definition of distribution.

We note that the original definition of cF, where c is a constant, is included in the general definition of product of distributions if c is considered as a constant-distribution.

We note also that when the Dirac δ and its derivatives $\delta^{(n)}$ are multiplied by an $\alpha(x)$ with derivatives $\alpha^{(n)}(x)$ of all orders, we obtain:

$$\alpha(x)\delta = \alpha(0)\delta,$$

$$\alpha(x)\delta = (\alpha\delta)' - \alpha'\delta = \alpha(0)\delta' - \alpha'(0)\delta,$$

and in general,

$$\alpha(x)\delta^{(n)} = \alpha(0)\delta^{(n)} - n\alpha'(0)\delta^{(n-1)} + \binom{n}{2}\alpha''(0)\delta^{(n-2)}$$
$$+ \ldots + (-1)^n\alpha^{(n)}(0)\delta.$$

§5. THE ORDER CLASSIFICATION OF DISTRIBUTIONS

THE system of c.l.f.'s on a given closed, finite interval includes, of course, every summable $f(x)$ and all its derivatives. We shall show in this section that *there are no other c.l.f.'s*, that is, every c.l.f. is either $f(\phi)$ or $f^{(n)}(\phi)$ for some suitable summable $f(x)$ and some finite n. (By using a higher n we can prove this with $f(x)$ continuous or even absolutely continuous.)

Let F be a c.l.f. on (a, b). F will be said to have *finite order* on (a, b) either if it can be identified with a summable point-function $f(x)$ or if, for some finite r, F is the rth order derivative of some such $f(x)$: $F = f^{(r)}$. The smallest possible r will be called the order of F.[7]

Clearly if F can be identified with a summable $f(x)$ then it has order 0 and its derivative F' has order either 0 or 1. If F has order r greater than 0 its derivative has order $r + 1$ precisely.

If F has order r and $s \geqslant r$ then for some f which depends on F and s, $F = f^{(s)}$; if $s > r$, f is absolutely continuous; if $s = 0$, f is uniquely determined to within Lebesgue equivalence, but if $s > 0$, f is determined only to within an additive (arbitrary) polynomial in x of degree $s - 1$.

We wish to show that F must have finite order. The definition of c.l.f. requires that $F(\phi_m)$ shall converge to $F(\phi)$ whenever all ϕ, ϕ_m are in $S_{(a,b)}$ and, for each $n \geqslant 0$, the $\phi_m^{(n)}(x)$ converge uniformly to $\phi^{(n)}(x)$. We shall now show that this implies the following apparently stronger condition. For some finite r which depends on F,

(C_r) $F(\phi_m)$ converges to $F(\phi)$ whenever all ϕ, ϕ_m are in $S_{(a,b)}$ and, for all n with $0 \leqslant n \leqslant r$, the $\phi_m^{(n)}(x)$ converge uniformly to $\phi^{(n)}(x)$.

Indeed, suppose if possible, that (C_r) is false for every r. We can then define a sequence of testing functions ϕ_m such that for each m,

(i) $|\phi_m^{(n)}| < 2^{-m}$ for all $n \leqslant m$,

(ii) $F(\phi_m) > 1$.

(We are using the notation $|\phi| = \max\,[|\phi(x)|;\, a \leqslant x \leqslant b]$.) Then for every n, the $\phi_m^{(n)}(x)$ converge uniformly to the zero function and since F is a c.l.f.,

[7]In *Théorie des Distributions* I, p. 25, a slightly different definition is used: the order of F is defined there to be the least r for which $F = F_0^{(r)}$ with F_0 a Stieltjes measure.

this should imply that $F(\phi_m) \to 0$. But this contradicts (ii) above and so (C_r) can not be false for every r, that is, (C_r) holds for some finite r.

But for $n < r$,

$$\phi^{(n)}(x) = \int_a^x \frac{(x-t)^{r-n-1}}{(r-n-1)!} \phi^{(r)}(t)dt$$

so that $|\phi^{(n)}| \leqslant K|\phi^{(r)}|$ for all $n < r$, for some finite K which depends only on r, a and b. It follows that the condition (C_r) is equivalent to the condition

(B_r) $F(\phi_m)$ converges to $F(\phi)$ whenever all ϕ, ϕ_m are in $S_{(a,b)}$ and the $\phi_m{}^{(r)}(x)$ converge uniformly to $\phi^{(r)}(x)$.

We now show that the condition (B_r) implies the condition

(A_r) $|F(\phi)| \leqslant |F|_r|\phi^{(r)}|$

for all ϕ in $S_{(a,b)}$ with $|F|_r$ a finite constant (we shall let $|F|_r$ actually denote the smallest possible such constant).

Indeed if (A_r) were false we could define a sequence ϕ_m with $F(\phi_m) > m|\phi_m{}^{(r)}|$. Then the functions $\mu_m(x) = |\phi_m{}^{(r)}|^{-1}m^{-1}\phi_m(x)$ would be in $S_{(a,b)}$ and the $\mu_m{}^{(r)}(x)$ would converge uniformly to zero since $|\mu_m{}^{(r)}| = m^{-1}$. But $F(\mu_m) > 1$, contradicting (B_r). Thus (A_r) cannot be false.

To summarize our results: if F is a c.l.f. on (a, b) there is a finite r for which $|F|_r < \infty$ and

$$|F(\phi)| \leqslant |F|_r|\phi^{(r)}|$$

for all ϕ in $S_{(a,b)}$.

Now we construct a new functional L, which is defined for functions which can be put in the form $\phi^{(r)}$, by the formula $L(\phi^{(r)}) = F(\phi)$. The functional L is, by the preceding paragraph, a bounded linear functional on the linear space of functions of the form $\phi^{(r)}$ with norm $|\phi^{(r)}|$. The Hahn-Banach procedure[8] can be used to extend this functional L to all continuous functions on (a, b) without increasing the bound of L. Then the representation theorem of F. Riesz[9] applies and shows that

$$L(\phi^{(r)}) = \int_a^b \phi^{(r)}(x)d\psi(x)$$

for some $\psi(x)$ of total variation equal to $|F|_r$ and $\psi(x)$ can be assumed to satisfy $|\psi(x)| \leqslant |F|_r$. Thus

[8]See S. Banach, *Théorie des Opérations Linéaires*, Warsaw, 1932, Théorème 2, p. 55.
[9]See *Théorie des Opérations Linéaires*, pp. 59-61.

$$F(\phi) = L(\phi^{(r)}) = \int_a^b \phi^{(r)}(x)d\psi(x)$$

$$= -\int_a^b \psi(x)\phi^{(r+1)}(x)dx$$

so that $F = f^{(r+1)}$ where $f = (-1)^r\psi$, showing that F is indeed of finite order. If we replace f by one of its indefinite integrals we can write $F = f^{(r+2)}$ with $f(x)$ absolutely continuous.

For use in the next section we emphasize that if $|F|_r$ is finite then $F = f^{(r+1)}$ with $f(x)$ bounded, in fact $|f| \leqslant |F|_r$. The converse to this is false; however $|F|_r$ is certainly finite if F is of order r since then

$$|F(\phi)| = |(-1)^r\int_a^b f(x)\phi^{(r)}(x)dx|$$

$$\leqslant \left(\int_a^b |f(x)|dx\right)|\phi^{(r)}|.$$

Thus every c.l.f. F on a finite closed interval can be expressed as a suitable derivative of a summable point-function, and our system of c.l.f.'s is extensive enough (but no more) to permit indefinite derivation of the summable point-functions.

(This suggests an abstract, but equivalent formulation of the system of c.l.f.'s as the system of all symbols $f^{(n)}$ using all summable $f(x)$ and all non-negative integers n, and imposing obvious rules of addition, identification and so on.)

If a distribution F on an open interval I is considered, then the above considerations apply to every $F_{(a,b)}$. We write $F_{(a,b)} = f^{(r)}$ on (a, b) and speak of F as having order r on (a, b). However, the order of F on (a, b) may be unbounded when a and b vary.

A simple example of this is the distribution

$$F(\phi) = \Sigma_{m=1}^{\infty}\phi^{(m)}(m)$$

on the interval $-\infty < x < \infty$. The sum on the right has only a finite number of non-zero addends for every testing function and the order of this F on an interval (a, b) is the largest integer n for which $a < n < b$.

§6. Continuity and convergence properties of distributions

Consider first c.l.f.'s on an interval (a, b).

A set of F_a will be called bounded if, for each ϕ in $S_{(a,b)}$, the numbers $F_a(\phi)$ are bounded, that is

$$|F_a(\phi)| \leqslant M$$

where M is a finite constant which depends, of course, on ϕ, $M = M(\phi)$.

We shall say that F_m form a convergent sequence if, for each ϕ, the numbers $F_m(\phi)$ form a convergent sequence, and that the F_m converge to F as limit if, for each ϕ, the $F_m(\phi)$ converge to $F(\phi)$ as limit. Obviously, if the F_m converge to a limit, then the F_m form a convergent sequence; conversely, if the F_m do form a convergent sequence, we define F by $F(\phi) = \lim_{m\to\infty} F_m(\phi)$; as we shall show later in this section, F is then a c.l.f. and the F_m converge to F as limit.

Finally, we shall say that the series $\Sigma_{m=1}^{\infty} F_m$ is convergent and has sum F if $\Sigma_{m=1}^{N} F_m$ converges to F as N becomes infinite.

We shall now show that if the F_a are bounded there must be a finite r such that

$$|F_a(\phi)| \leqslant K|\phi^{(r)}|$$

for all a, ϕ for some finite constant K, that is, the $|F_a|_r$ are bounded. Indeed, if this were false for every r, we could, by induction on p select a sequence of ϕ_p and F_{a_p} such that, for each p,

$$\text{(i) } |\phi_p^{(n)}| \quad < 2^{-p} \text{ for all } n \leqslant p,$$

$$\text{(ii) } F_{a_p}(\phi_p) \ > p + 1 + \Sigma_{m=1}^{p-1} M(\phi_m),$$

$$\text{(iii) } F_{a_p}(\phi_m) < 2^{-m} \text{ for all } m > p.$$

For let $i = 0, 1, \ldots, p-1$ and suppose the ϕ_i, F_{a_i} have been selected to satisfy (i), (ii), (iii) in so far as they are involved in these conditions. Since each F_{a_i} has $|F_{a_i}|_{r_i}$ finite for some r_i we will have $|F_{a_i}(\phi_p)| < 2^{-p}$ for all i if ϕ_p satisfies the p conditions

$$|\phi_p^{(r_i)}| < 2^{-p}|F_{a_i}|_{r_i}^{-1} \qquad i = 0, 1, \ldots, p-1.$$

These p conditions as well as (i) can be included in a single condition

$$|\phi_p^{(r)}| < K_1$$

with a suitable finite K_1 and $r = \max(r_0, \ldots, r_{p-1}, p)$. Since we are assuming that the $|F_a|_r$ are unbounded for every r, there must be an F_{a_p} and a ϕ_p such that ϕ_p satisfies this condition and $F_{a_p}(\phi_p)$ satisfies (ii). Then the ϕ_i, F_{a_i} with $i = 0, 1, \ldots, p$ will satisfy (i), (ii), (iii) in so far as they are involved in these conditions.

Then $\Sigma_{m=1}^{\infty}\phi_m(x)$ would be a testing-function $\phi(x)$ for which

$$F_{a_p}(\phi) = F_{a_p}(\Sigma_{m=1}^{\infty}\phi_m) = \lim_{N\to\infty} F_{a_p}(\Sigma_{m=1}^{N}\phi_m) > p$$

and hence, for this ϕ, the $F_a(\phi)$ would not be bounded. This contradiction shows that the F_a are bounded if and only if $|F_a(\phi)| \leqslant K\,|\phi^{(r)}|$ for some fixed K and r, for all a and all testing-functions.

Using the result emphasized in the preceding section we can conclude that c.l.f.'s F_a are bounded if and only if they can be expressed as $F_a = f_a^{(r)}$ for some common r, with $|f_a|$ bounded.

As a corollary, we deduce that if F_m form a convergent sequence, then $F(\phi) = \lim_{m\to\infty} F_m(\phi)$ is a c.l.f. and the F_m converge to F; for the convergence of the F_m implies that they are bounded, hence $|F_m|_r \leqslant K$ for some r and some fixed finite K; hence

$$|F(\phi)| = \lim_{m\to\infty} \quad |F_m(\phi)| \leqslant K|\phi^{(r)}|$$

which implies that F is a c.l.f.

If the F_m can be expressed as $F_m = f_m{}^{(r)}$ with the $f_m(x)$ converging uniformly, then

$$F_m(\phi) = (-1)^r \int_a^b f_m(x)\phi^{(r)}(x)dx \to (-1)^r \int_a^b (\lim_{m\to\infty} f_m(x))\phi^{(r)}(x)dx$$

so that the F_m do converge and $\lim_{m\to\infty} F_m = \left\{\lim_{m\to\infty} f_m(x)\right\}^{(r)}$. We now show that the converse also holds, that is, if the F_m are convergent then for some r, $F_m = f_m{}^{(r)}$ with $f_m(x)$ converging uniformly (the f_m can actually be chosen to be continuous) and $\lim_{m\to\infty} F_m = \{\lim_{m\to\infty} f_m(x)\}^{(r)}$.

Suppose therefore that the F_m are a convergent sequence so that the $F_m(\phi)$ converge for every ϕ to a limit $F(\phi)$. Then the F, F_m are bounded and hence

$$F_m = f_m{}^{(r)}, \quad F = f^{(r)}$$

for some r and suitable $f(x)$, $f_m(x)$ with $|f|$, $|f_m|$ bounded. By replacing each of $f(x)$, $f_m(x)$ by its integral from a to x and using r in place of the former $r + 1$ we can suppose that $F = f^{(r)}$, $F_m = f_m{}^{(r)}$ with the $f(x)$, $f_m(x)$ *equicontinuous* on (a, b). We can of course also write $F_m = (f_m + P_m)^{(r)}$ where the $f_m(x)$ are as before and the $P_m(x)$ are arbitrary polynomials of degree less than r. We shall show in a separate lemma below that these polynomials can be chosen so as to give: $f_m(x) + P_m(x)$ converges uniformly to $f(x)$. This will prove the theorem: F_m converges if and only if $F_m = f_m{}^{(r)}$ for some r and uniformly convergent continuous $f_m(x)$, and then $F = \lim_{m\to\infty} F_m = (\lim_{m\to\infty} f_m(x))^{(r)}$.

It remains therefore to prove the following lemma, with $g_m(x)$ in place of our former $f_m(x) - f(x)$.

Lemma. *If $g_m(x)$ are equicontinuous in x on (a, b) and if for some fixed r and every testing function ϕ, $\int_a^b g_m(x)\phi^{(r)}(x)dx$ converges to zero as m becomes infinite, then there are polynomials $P_m(x)$ of degree less than r such that $g_m(x) + P_m(x)$ converges uniformly to zero as m becomes infinite.*

We may suppose that the $g_m(x)$ are real-valued by considering real and imaginary parts separately if they are complex-valued.

We first prove the lemma for the case $r = 0$. Let ϵ be an arbitrary positive number and $\delta(\epsilon)$ be such that for all m, $|g_m(x) - g_m(y)| < \epsilon$ whenever

$|x - y| < \delta(\epsilon)$. Let (a, b) be covered by a finite number of intervals $I_1, \ldots, I_t$ each of length less than $\delta(\epsilon)$. Suppose that for some q and some $I_p = (c, d)$ say, $g_q(x) \geqslant \epsilon$ for all x in I_p. Then

$$\int_a^b g_q(x)\phi_{c,d}(x)dx \geqslant \epsilon \int_c^d \phi_{c,d}(x)dx.$$

($\phi_{c,d}$ was defined in §2.) Since $\int_a^b g_m(x)\phi_{c,d}(x)dx \to 0$ as $m \to \infty$ it is impossible for this to be true for an infinite number of q for the same I_p. Since there are only a finite number of such I_p there is an m_0 such that for $m \geqslant m_0$, $g_m(x) < \epsilon$ for at least one x_p in each I_p. But every x in (a, b) lies in some I_p, so

$$g_m(x) = g_m(x) - g_m(x_p) + g_m(x_p),$$

i.e., for all x in (a, b) and for all $m \geqslant m_0$, $g_m(x) < 2\epsilon$. Similarly we can obtain $-g_m(x) < 2\epsilon$ and hence $|g_m(x)| < 2\epsilon$ for all x, for all but a finite number of m. This means that the $g_m(x)$ converge uniformly to zero, proving the lemma for the case $r = 0$.

Now to prove the lemma for all r by induction. Suppose that $\theta(x)$ is a fixed function in $S_{(a,b)}$ with $\int_a^b\theta(x)dx = 1$ and use the expansion lemma of §3 to write $\phi(x) = a_0\theta(x) + \rho_1'(x)$ where $a_0 = \int_a^b\phi(x)dx$. Then

$$\begin{aligned}\int_a^b g_m(x)\phi^{(p)}(x)dx &= \int_a^b g_m(x)\{a_0\theta^{(p)}(x) + \rho_1^{(p+1)}(x)\}dx \\ &= c_m\int_a^b \phi(x)dx + \int_a^b g_m(x)\rho_1^{(p+1)}(x)dx\end{aligned}$$

where

$$c_m = \int_a^b g_m(x)\theta^{(p)}(x)dx$$

and depends on g_m but not on ϕ, and is bounded for all m. But repeated use of integration by parts shows that

$$\begin{aligned}c_m\int_a^b \phi(x)dx &= c_m\frac{(-1)^p}{p!}\int_a^b x^p\phi^{(p)}(x)dx \\ &= -b_m\int_a^b x^p\phi^{(p)}(x)dx\end{aligned}$$

where b_m is a constant bounded in m depending on g_m but not on ϕ. Hence we have the identity:

$$\int_a^b (g_m(x) + b_m x^p)\phi^{(p)}(x)dx = \int_a^b g_m(x)\rho_1^{(p+1)}(x)dx$$

and it is clear that if the lemma holds for $r = p$, it also holds for $r = p + 1$. Therefore it holds for all r.

We may now consider distributions on an open interval I. We shall say that the F_a are bounded if for each ϕ, the $F_a(\phi)$ are bounded and we shall say F_m converges to F if, for each ϕ, $F_m(\phi)$ converges to $F(\phi)$. Now our previous work enables us to draw relevant conclusions about the $F_{a(a,b)}$, $F_{(a,b)}$ for each (a, b) in I.

An important convergence-property of distributions is this: if F_m converges to F then F'_m converges to F'. This follows immediately from the fact that

$$F'(\phi) - F'_m(\phi) = - \{F(\phi') - F_m(\phi')\}.$$

Thus, if $F = \Sigma_{m=1}^{\infty} F_m$ then $F' = \Sigma_{m=1}^{\infty} F'_m$ so that term-by-term differentiation is valid without restriction for *distributions*. In particular, if $f(x)$ is the sum $\Sigma_{m=1}^{\infty} f_m(x)$ of a uniformly convergent series of functions $f_m(x)$ each of which is absolutely continuous, then $f(x)$ may not be absolutely continuous but it will be summable (even continuous) and hence have a distribution-derivative f' which will be the limit (distribution-limit) of $\Sigma_{m=1}^{N} f'_m(x)$ as N becomes infinite; if $f(x)$ happens to be absolutely continuous, then $\Sigma_{m=1}^{N} f'_m(x)$ will converge (as a distribution but not necessarily point-wise) to $f'(x)$.

Let us say that a distribution F_t, depending on a parameter t, converges to F as t approaches t_0, if, for each ϕ, $F_t(\phi)$ converges to $F(\phi)$ as t approaches t_0. Then convergence of F_t to F implies that of F'_t to F'.

Thus, starting from the function

$$f(x) = \lim_{t \to \infty} \int_1^t - \frac{\cos wx}{w^2} dw$$

we can differentiate twice, in the sense of distributions to obtain

$$f'' = \lim_{t \to \infty} \int_1^t \cos wx dw$$

so that we can give a meaning to $\int_1^{\infty} \cos wx dw$ as a distribution although the integral is not convergent in the usual sense.

Let us calculate the value of $2\int_0^{\infty} \cos 2\pi wx dw$, which, according to the preceding paragraph, will have a meaning as a distribution. We set

$$g_t(x) = \int_0^t 2 \cos 2\pi wx dw = \frac{\sin 2\pi tx}{\pi x}$$

and observe that

$$g_t(\phi) = \int_{-\infty}^{+\infty} \frac{\sin 2\pi tx}{\pi x} \phi(x) dx$$

converges to $\phi(0)$ as t becomes infinite (the well-known Dirichlet integral of Fourier series), so that g_t as a distribution converges to the Dirac δ. This gives the result

$$2\int_0^\infty \cos 2\pi wx dw = \delta.$$

Such formulae have been used classically in the theory of electricity, in symbolic calculations in wave mechanics, but without adequate explanation. As distribution-formulae they are, however, unambiguous and have rigorous validity.

If F_t is a distribution depending on a parameter t we can define $\frac{dF}{dt}$ (not to be confused with F') as that distribution F for which $F(\phi) = \frac{d}{dt} F_t(\phi)$ for all ϕ, if such an F exists. Similarly we can define $\int_{t_1}^{t_2} F_t dt$ to be F if $\int_{t_1}^{t_2} F_t(\phi)dt$ exists and equals $F(\phi)$ for all ϕ.

§7. Nullity-sets and supporting-sets

A distribution F, even though it be more general than a point-function, can be said to have "local" properties, that is, properties which concern a particular (but arbitrary) x and neighbourhoods of that x. We have already seen that for each x there is an interval (c, d) with $c < x < d$ such that F has finite order on (c, d).

We shall say that a particular x_0 in I is in the *nullity-set* of a distribution F on I if $F_{(c,d)} = 0$ for some $c < x_0 < d$. The complement with respect to I of the nullity-set of F will be called the *supporting-set* of F. Obviously the nullity-set is open and the supporting-set is closed (relative to the containing interval I).

If F happens to be identifiable with a *continuous* point-function $f(x)$ then the supporting-set of F will clearly be the closure of the set of x for which $f(x) \neq 0$, hence the closure of an open set; the nullity-set will be the interior of the set of x for which $f(x) = 0$, hence the interior of a closed set. But these statements are not valid for arbitrary locally summable (but discontinuous) $f(x)$.

The zero-distribution has a nullity-set which includes all x so that its supporting-set is the empty set. Conversely, the zero-distribution is the only distribution with these properties. For on any (a, b) contained in I, $F = f^{(r)}$ with $f(x)$ continuous on (a, b) so that

$$F(\phi) = (-1)^r \int_a^b f(x)\phi^{(r)}(x)dx$$

for all ϕ in $S_{(a,b)}$, and hence for all ϕ in $S_{(c,d)}$ with $a \leqslant c < d \leqslant b$. If some particular $F_{(c,d)} = 0$ the Lemma of the preceding §6 (with all g_m of the Lemma equal to f) implies that $f(x)$ is a polynomial of degree less than r on (c, d). Hence, by the Heine-Borel Theorem, (a, b) can be covered by a finite number of such (c, d) on each of which $f(x)$ is a polynomial of degree less than r. Thus $f(x)$ is a polynomial of degree less than r on (a, b) and $F(\phi) = 0$ for all ϕ in $S_{(a,b)}$. Therefore $F = 0$ as required.

This type of reasoning also shows that for a particular ϕ, $F(\phi)$ will certainly have the value 0 if the closure of the set of x for which $\phi(x) \neq 0$ is contained in the (open) nullity-set of F. It follows that the value of $F(\phi)$ depends only on the values of $\phi(x)$ in the neighbourhood of the supporting-set of F, that is, on the values of $\phi(x)$ in any open set containing the supporting-set.

The preceding statement cannot be sharpened to imply that $F(\phi)$ depends only on the values of $\phi(x)$ on the supporting-set of F; the derivative of the Dirac δ, namely δ', has the origin as the only point in its supporting-set yet $\delta'(\phi) = -\phi'(0)$ is not determined by the value of $\phi(0)$. But since on every (a, b) each F has the form $f^{(r)}$ we can show that for each (a, b) and each F there is a fixed r such that the value of $F(\phi)$, for ϕ in $S_{(a,b)}$, depends only on the values of $\phi(x)$, $\phi'(x)$, . . . , $\phi^{(r)}(x)$, on the supporting-set of F. Indeed, N, the nullity-set of F, is an open set and can be expressed as the set-union of a finite or countable number of disjoint open intervals $< c_n, d_n >$. Then

$$F(\phi) = (-1)^r \int_a^b f(x)\phi^{(r)}(x)dx$$

$$= \Sigma_n(-1)^r \int_{c_n}^{d_n} f(x)\phi^{(r)}(x)dx + (-1)^r \int_{I-N} f(x)\phi^{(r)}(x)dx.$$

On each $< c_n, d_n >$, $f(x)$ is a polynomial $P_n(x)$ of degree $r - 1$ at most, so that $\int_{c_n}^{d_n} f(x)\phi^{(r)}(x)dx$ when integrated, depends only on the values at c_n and d_n of $\phi(x)$, $\phi'(x)$, . . . , $\phi^{(r-1)}(x)$. Since all c_n, d_n are in the supporting-set of F and since the value of the integral of $f(x)\phi^{(r)}(x)$ over $I - N$, the supporting-set of F, depends obviously only on the values of $\phi^{(r)}$ on the supporting-set of F, the desired conclusion can be drawn.

A corollary of the previous paragraph is this: a distribution has a supporting-set consisting of one point only, the origin, if and only if the distribution is a finite linear combination of the Dirac δ and its derivatives:

$$F = \Sigma_{p=0}^{N} c_p \delta^{(p)}.$$

For any fixed x_0, let δ_{x_0} denote the distribution

$$\delta_{x_0}(\phi) = \phi(x_0)$$

so that the δ of our previous notation coincides with δ_0. From now on we shall call any such distributions δ_{x_0} a Dirac delta. Then it follows, as above, that a distribution on I has a supporting-set consisting of isolated points if and only if the distribution is a finite or infinite linear combination of Dirac deltas and their derivatives such that for each (a, b) in I, only a finite number of these deltas and their derivatives are taken at points in (a, b).

It is easily seen that the supporting-set of F contains the supporting-set of F' and an analysis of the non-increasing family of supporting-sets of $F^{(n)}$ gives some information about the local structure of F.

For use later in this section we require the following observations.

If $a < c < d < b$ there is a continuous function $\alpha(x)$ with derivatives $\alpha^{(n)}(x)$ of all orders such that $\alpha(x) = 0$ for $x \leqslant c$ and $\alpha(x) = 1$ for $x \geqslant d$; such a function can be obtained by defining $\alpha(x)$ for $c < x < d$ as

$$\left\{\int_c^d \phi_{c,d}(t)dt\right\}^{-1}\int_c^x \phi_{c,d}(t)dt$$

($\phi_{c,d}$ was defined in §2). Hence every ϕ in $S_{(a,b)}$ can be put in the form $\phi_1 + \phi_2$ with ϕ_1 in $S_{(a,d)}$ and ϕ_2 in $S_{(c,b)}$; put $\phi_1 = (1 - \alpha)\phi$ and $\phi_2 = \alpha\phi$. Repeated applications of this observation show that if (a, b) can be covered by a finite number of open intervals $I_p = (c_p, d_p)$, $p = 1, \ldots, N$, then every ϕ in $S_{(a,b)}$ can be expressed as a sum $\Sigma_{p=1}^{N}\phi_p$ with ϕ_p in $S_{(c_p,d_p)}$. Finally, if each x_0 in (a, b) is covered by at least one of a family of open intervals I_α, then (a, b) is covered by a finite number of such I_α by the Heine-Borel theorem. This leads to the partition theorem: A finite number of (c, d) can be selected, each an I_α, so that every ϕ can be expressed as a sum of ϕ_p, with each ϕ_p in one of the $S_{(c,d)}$. Moreover this can be done in such a way that the condition on a sequence of ϕ that they and all their nth order derivatives converge uniformly to zero is equivalent to the same condition for their ϕ_p for each p.

Now suppose that $T(\phi)$, not assumed to be a distribution, is defined for certain ϕ as follows: for a family of (c, d) contained in I, $T(\phi)$ is defined for all ϕ in $S_{(c,d)}$ and, when restricted to $S_{(c,d)}$, is a c.l.f. $T_{(c,d)}$ on (c, d); suppose too that each x_0 in I is covered by the interior of at least one (c, d). In other words, suppose that $T(\phi)$ defines a distribution *locally*. Then we assert: there is one and only one distribution F on I which extends T, that is, for which $F_{(c,d)} = T_{(c,d)}$ for all such (c, d). For an arbitrary ϕ can be expressed as the sum of a finite number of ϕ_p, with ϕ_p in some $S_{(c,d)}$. Then $T\phi_p$ is defined for each p and F, if it exists, must satisfy $F(\phi) = \Sigma_p T(\phi_p)$ (this gives a simple proof that if F is a distribution with nullity-set which includes all x, then F must be the zero-distribution). On the other hand, this actually defines $F(\phi)$ uniquely and this F is a distribution with the properties stated above, as follows from the preceding paragraph.

The preceding observations enable us to generalize our definition of product of two distributions. If F_1, F_2 are two distributions on I we shall say that their product is defined locally at x_0 if $F_{1(c,d)}F_{2(c,d)}$ is defined, in accordance with our previous definition, for some (c, d) with interior covering x_0. If the product of F_1 and F_2 is defined locally at every x, there is one and only one distribution F on I such that $F_{(c,d)} = F_{1(c,d)}F_{2(c,d)}$ for all such (c, d); we define this F to be the product F_1F_2. In order that the product of F_1 and F_2 be defined locally at x_0 it is easily seen that a necessary (but not sufficient) condition is this: x_0 must not be an isolated point in the supporting-sets of both F_1 and F_2. Thus F_1F_2 is not defined if there are any common isolated points in the supporting-sets; for example, the product of δ_0 by itself is not defined. On the other hand F_1F_2 will certainly be defined and will be the zero-distribution if the supporting-sets have no points in common (that this sufficient condition is not necessary is shown by the fact that $x\delta_0$ is the zero-distribution).

§8. DIFFERENTIAL EQUATIONS INVOLVING DISTRIBUTIONS

CONSIDER the differential equation

$$G^{(m)} + A_{m-1}G^{(m-1)} + \ldots + A_0G = F$$

under the assumptions that each A_{m-s} is a bounded function of x on a closed finite interval (a, b) and that F is a function of x summable on (a, b). A point-function $G(x)$ is a *solution* of the differential equation if $G(x), \ldots, G^{(m-1)}(x)$ are absolutely continuous and the $G(x), \ldots, G^{(m)}(x)$ satisfy the given equation for almost all x. All such solutions can be found by the classical method of successive substitutions as follows:

Let $H^{[-1]}(x)$ denote $\int_a^x H(t)dt$ for any summable $H(x)$, so that

$$H^{[-s]}(x) = \int_a^x \frac{(x-t)^{s-1}}{(s-1)!} H(t)dt.$$

Let ΔH denote $-[A_{m-1}(x)H^{(m-1)}(x) + \ldots + A_0(x)H(x)]$ for any $H(x)$ such that $H(x), \ldots, H^{(m-1)}(x)$ exist and are absolutely continuous. Then set

$$G_0(x) = F^{[-m]}(x)$$

and

$$G_{k+1}(x) = F^{[-m]}(x) + (\Delta G_k)^{[-m]}(x) \qquad \text{for } k \geqslant 0.$$

The series

$$G_0(x) + \Sigma_{k=0}^{\infty}(G_{k+1}(x) - G_k(x))$$

will converge uniformly to a sum $G(x)$ and the first, second $\ldots$, mth derived series will converge uniformly to $G'(x), \ldots, G^{(m)}(x)$ respectively. From this it can be shown that $G(x)$ is a solution of the given differential equation, to be called *a particular integral*. From the construction it also follows that if, for any p, all coefficients $A_{m-s}(x)$ have $A_{m-s}(x), \ldots, A_{m-s}{}^{(p)}(x)$ absolutely continuous and $A_{m-s}{}^{(p+1)}(x)$ bounded, and $F(x)$ has $F(x)$, $F'(x)$, $\ldots$, $F^{(p)}(x)$ absolutely continuous, then $G(x)$ has $G(x), G'(x), \ldots, G^{(m+p)}(x)$ all absolutely continuous. In particular, if the $A_{m-s}(x)$ and $F(x)$ are indefinitely differentiable then $G(x)$ is indefinitely differentiable.

Now m linearly independent solutions of the homogeneous equation (replacing F by 0) can be obtained by choosing r to be in turn $0, 1, \ldots, m-1$ and setting:

$$G_0(x) = \frac{x^r}{r!},$$

$$G_{k+1}(x) = (\Delta G_k)^{[-m]}(x),$$

$$G(x) = G_0(x) + \Sigma_{k=0}^{\infty}(G_{k+1}(x) - G_k(x)).$$

Without danger of confusion, we now adopt the notation $G_p(x)$ for the particular integral and $G_0(x), \ldots, G_{m-1}(x)$ for the solutions of the homogeneous equation found above. Then for any constants $c_0, \ldots, c_{m-1}$, the function

$$G(x) = G_p(x) + c_0G_0(x) + \ldots + c_{m-1}G_{m-1}(x)$$

will be a solution of the given differential equation, and G_p will have differentiability properties, as explained above, depending on those of F and the A_{m-s}, while $G_0, \ldots, G_{m-1}$ will have differentiability properties depending on those of the A_{m-s} only. As can be shown easily, there are no other point-function solutions of the given differential equation.

We now regard the given differential equation as an equation in an (unknown) c.l.f. G. The point-function solutions previously found are also c.l.f.'s. *We shall show that there are no other c.l.f. solutions.* Indeed, if the order of $G^{(m)}$ were greater than zero, it would be greater than the orders of each of $G^{(m-1)}, \ldots, G, F$ and hence greater than the order of

$$F - (A_{m-1}G^{(m-1)} + \ldots + A_0G)$$

which is supposed to be equal to $G^{(m)}$. This contradiction shows that the order of $G^{(m)}$ must be zero, so that $G^{(m)}$ is a summable point-function and hence G is a point-function solution as defined above.

The above technique of solving the differential equation can be applied if F is an arbitrarily assigned c.l.f. provided that the A_{m-s} are restricted to be bounded point-functions satisfying another condition. Let F^{-1} denote any any primitive of F, and by induction, F^{-s} any primitive of $F^{-(s-1)}$; of course the F^{-s} are defined only to within an additive (arbitrary) polynomial of degree $s-1$. Now we require that $A_{m-s}F^{-s}$ be defined (as a product of two c.l.f.'s for $s = 1, \ldots, m$ (this condition is certainly satisfied if the $A_{m-s}(x)$ happen to be all indefinitely differentiable).

Under the above assumptions if the order of F is greater than zero, we rewrite the differential equation as:

$$(G - F^{-m})^{(m)} + A_{m-1}(G - F^{-m})^{(m-1)} + \ldots + A_0(G - F^{-m}) = F_1$$

where

$$F_1 = -[A_{m-1}F^{-1} + A_{m-2}F^{-2} + \ldots + A_0F^{-m}].$$

Now we have an equation in $G - F^{-m}$ in which the right side F_1 is of order less than that of F. By repeated reductions of the order of the right side we finally obtain a differential equation in which the right side is a summable point-function. The use of ordinary point-function integrals can then be combined with the method of successive substitutions to yield the theorem: there is a c.l.f. G_p and m linearly independent point-functions $G_0, \ldots, G_{m-1}$ such that

$$G = G_p + c_0G_0 + \ldots + c_{m-1}G_{m-1}$$

is a c.l.f.-solution of the given differential equation for arbitrary $c_0, \ldots, c_{m-1}$, and there are no other c.l.f.-solutions. The order of $G_p^{(m)}$ is precisely the order of F and the $G_0(x), \ldots, G_{m-1}(x)$, as point-function solutions of the homogeneous equation, have differentiability properties depending on those of the $A_{m-s}(x)$.

This discussion of c.l.f.-solutions on each (a, b) leads without difficulty to the corresponding distribution-solutions of the differential equation on an open interval I.

The more general equation

$$A_m(x)G^{(m)} + A_{m-1}(x)G^{(m-1)} + \ldots + A_0(x)G = F$$

can be reduced to the case $A_m = 1$ providing that the coefficient $A_m(x)$ can be *factored out*. Then the system of solutions is as described above. However, in the singular case, where $A_m(x)$ vanishes at certain points, quite different results may be obtained. Under some circumstances the homogeneous equation may have no solutions other than the trivial one—the identically zero distribution. Under other circumstances, the solutions may depend on a number of arbitrary parameters greater than the order of the equation. For example, consider the equation

$$xG' + G = 0.$$

This can be written as $(xG)' = 0$ and the solutions must satisfy $xG = k$ for some constant k. This seems to give a one-parameter family of solutions $G = \frac{k}{x}$. These are indeed solutions in a certain sense (they are not absolutely continuous over any interval including the origin) but they are not locally summable functions and hence cannot be identified with distributions. From the point of view of distributions, the equation $xG = k$ does have solutions, however, for instance

$$G(\phi) = \text{pv.} \int_a^b \frac{k\phi(x)}{x}\,dx$$

where pv. means Cauchy principal value. If we denote this solution by $k\left(\text{pv.}\frac{1}{x}\right)$ then all other solutions differ from it by a solution of the equation $xG_1 = 0$: such a G_1 must have supporting-set consisting of one point, the origin, and hence has the form

$$G_1 = \Sigma_{r=0}^{N} c_r \delta_0^{(r)}.$$

Then

$$xG_1 = \Sigma_{r=0}^{N} c_r (x\delta_0^{(r)}) = \Sigma_{r=0}^{N} c_r (-r)\delta_0^{(r-1)} = 0$$

if and only if $c_r = 0$ for $r > 0$. Thus $G_1 = c_0\delta_0$ is the set of all such solutions G_1. Hence the solutions of the equation

$$xG' + G = 0$$

consists precisely of all distributions $G = k_1\left(\text{pv.}\frac{1}{x}\right) + k_2\delta_0$ and depends on two independent parameters.

§9. Distributions in several variables; definitions and examples

The preceding theory can be extended to distributions in n variables where n is any positive integer. In this section we give definitions and some examples; in the next section we will sketch some of the theory.

Let x now denote an n-tuple of real numbers: $x = (x_i) = (x_1, \ldots, x_n)$ which may be thought of as a point in n-dimensional space. $x \leqslant y$ will mean $x_i \leqslant y_i$ for all i. If $a = (a_i)$, $b = (b_i)$ are arbitrary but fixed, with $a_i < b_i$ for all i we let R denote the closed rectangle $a \leqslant x \leqslant b$. S_R will denote the set of all continuous functions $\phi(x) = \phi(x_1, \ldots, x_n)$ which possess continuous partial derivatives of all orders:

$$\phi^{(p)}(x) = \phi^{(p_1, \ldots, p_n)}(x_1, \ldots, x_n)$$
$$= \frac{\partial^{p_1 + \cdots + p_n}}{(\partial x_1)^{p_1} \ldots (\partial x_n)^{p_n}} \phi(x_1, \ldots, x_n)$$

and which, together with all $\phi^{(p)}(x)$, vanish on the boundary of R and outside R. A linear functional $F(\phi)$ defined for all ϕ in S_R will be called a c.l.f. on R if: whenever ϕ, ϕ_m are in S_R and for each p, $\phi_m^{(p)}(x)$ converges uniformly to $\phi^{(p)}(x)$, then $F(\phi_m)$ converges to $F(\phi)$.

If Ω is any bounded or unbounded open rectangle in n-dimensional space (for example, all n-dimensional space), we define F to be a distribution on Ω if for each R contained in Ω, $F(\phi)$ is defined for ϕ in S_R and ,when restricted to such ϕ, is a c.l.f., to be denoted by F_R.

For any distribution F on Ω and any p, we define a partial distribution-derivative on Ω by the formula:

$$F^{(p)}(\phi) = (-1)^{p_1 + \cdots + p_n} F(\phi^{(p)}).$$

Then every distribution has partial derivatives of all orders and the result of successive derivations is always independent of the order of derivation.

A distribution F on Ω will be identified with a point-function $f(x)$ if, for each R contained in Ω, $f(x)$ is summable on R and

$$F(\phi) = \int f(x)\phi(x)dx = \int \ldots \int f(x_1, \ldots, x_n)\phi(x_1, \ldots, x_n)dx_1 \ldots dx_n$$

for each ϕ in S_R, the integration being taken over R or, what is equivalent, over all n-dimensional space.

Identification of certain distributions with Stieltjes-measures (in one or more of the x_i) can also be made. To take one example, suppose $\psi(x_1, x_2, x_3, \ldots, x_n)$ is a real-valued function and suppose that on each R contained in Ω the following holds: for almost all fixed $(x_3, \ldots, x_n)$, the function ψ considered as a function of x_1, x_2 has finite variations, $v_+(x_3, \ldots, x_n)$, $v_-(x_3, \ldots, x_n)$ which are summable over R. Then ψ determines a distribution F on Ω by the formula:

$$F(\phi) = \int \ldots \int dx_3 \ldots dx_n (\iint \phi(x_1, \ldots, x_n)d_{x_1 x_2}\psi(x_1, x_2, \ldots, x_n)).$$

A set of distributions F_a will be said to be bounded if, for each fixed ϕ, the $F_a(\phi)$ are bounded. F_m will be said to converge as m becomes infinite if $F_m(\phi)$ converges as m becomes infinite, for each fixed ϕ; in which case, as we shall prove in the next section, the limiting value of $F_m(\phi)$ will determine a limit distribution $F(\phi)$. Similarly F_t will be said to converge to F as $t \rightarrow t_0$ if $F_t(\phi) \rightarrow F(\phi)$ for each fixed ϕ. A series $\Sigma_{m=1}^{\infty} F_m$ will be said to have sum F if $\Sigma_{m=1}^{h} F_m$ converges to F as $h \rightarrow \infty$. Continuity, respectively convergence, of distributions implies that of their distribution-derivatives of all orders, as is easily seen, so that term-by-term differentiation is always valid.

A point x_0 will be said to be in the nullity-set of F if for some R which covers x_0, F_R is the identically zero c.l.f. The complement of the nullity-set of F will be called the supporting-set of F.

The sum of two distributions and a constant times a distribution are defined in the obvious way.

In the next section we will develop some of the theory of distributions in n variables. We conclude this section with examples.

For fixed x_0 let F be the distribution on all n-dimensional space: $F(\phi) = \phi(x_0)$. This is a generalization of the Dirac delta-function. It can be identified with the n-dimensional Stieltjes-measure $d_{x_1 \ldots x_n} Y(x_1, \ldots, x_n)$ where $Y(x)$ is the Heaviside function:

$$Y(x) = 1 \text{ for } x_0 \leqslant x \text{ and } Y(x) = 0 \text{ for all other } x.$$

In our terminology, this Dirac delta-function, δ_{x_0}, is the pth derivative of $Y(x)$ where $p = (1, 1, \ldots, 1)$. In terms of physical concepts such as mass and charge distribution, this delta-distribution corresponds to a unit of mass or charge concentrated at one point, namely x_0.

If we use the same $Y(x)$ but take the pth derivative for any $p = (p_i)$ with $p_i = 0$ or 1 for each i, we obtain a variety of mixed delta-distributions: thus $\frac{\partial Y}{\partial x_1}$ is the distribution

$$F(\phi) = \int \ldots \int \phi(x_{01}, x_2, \ldots, x_n) dx_2 \ldots dx_n,$$

the integration being taken over the $n - 1$ dimensional quadrant $x_1 = x_{01}$, $x_i \geqslant x_{0i}$ for $i > 1$. In physical terms, this corresponds to an $n - 1$ dimensional surface-distribution with unit surface-density on the quadrant.

The Heaviside function defined above has the value 1 on an n-dimensional quadrant and vanishes outside the quadrant. We now generalize by replacing the quadrant by a rectangle or sphere or any other n-dimensional set J which has a hypersurface H sufficiently smooth to permit calculations using integration by parts. Thus we now set $Y(x) = 1$ for x inside J or on the hypersurface H of J and we set $Y(x) = 0$ for all other x. Then $\frac{\partial Y}{\partial x_1}$, as a distribution, corresponds to a surface-distribution of mass or charge on H with surface density at a point of H equal to $\cos \theta_1$ (θ_i shall denote the angle between the inner normal to H and the positive x_i axis).

More generally, let $f(x)$ be a point-function which is continuous and has continuous partial derivatives when considered only within a closed set J (that is, including its hypersurface H). Suppose, however, that $f(x)$ vanishes outside J. Then f determines a distribution F. Calculation shows that the derivative-distribution $\frac{\partial F}{\partial x_1}$ is made up of two terms. The first term is a distribution which can be identified with the "ordinary" derivative $\frac{\partial f}{\partial x_1}$ which is supposed to exist at all x with possible exceptions for x on H; the second term is a distribution corresponding to a surface-distribution on H whose density at a point of H is the value of f at that point multiplied by $\cos \theta_1$. The distribution $\frac{\partial^2 F}{(\partial x_1)^2}$ consists of three terms. The first term corresponds to an n-dimensional volume-distribution over J with density $\frac{\partial^2 f}{(\partial x_1)^2}$: the second term corresponds to a hyper-surface distribution on H with density $\frac{\partial f}{\partial x_1} \cos \theta_1$: the third term corresponds to a layer of doublets or dipoles all parallel to the x_1 axis, and carried by H with surface-density $- f(x) \cos \theta_1$. The Laplacian distribution of F, namely

$$\frac{\partial^2 F}{(\partial x_1)^2} + \frac{\partial^2 F}{(\partial x_2)^2} + \dots + \frac{\partial^2 F}{(\partial x_n)^2}$$

written ΔF, also consists of three terms. The first term corresponds to a volume mass distribution over J, identifiable with the ordinary Laplacian of f: the second term corresponds to a surface-distribution H with density $\frac{df}{dn}$ $\left(\frac{d}{dn}\right.$ indicating derivative along the inner normal$\left.\right)$: and the third term corresponds to a double layer (as in Potential Theory), that is, a layer of doublets carried by H, all normal to H, with surface density $-f(x)$. In mathematical symbols,

$$(\Delta F)(\phi) = F(\Delta \phi) = \int_J (f)(\Delta \phi) dx$$
$$= \int_J (\Delta f)(\phi) dx + \int_H \left(\frac{df}{dn}\right)(\phi) dS - \int_H (f)\left(\frac{d\phi}{dn}\right) dS.$$

This formula is the well-known Green's Formula in Potential Theory given here with a distribution "interpretation." In the same way the formulae of Ostrogradski (Gauss), Riemann, Green, and Stokes can be given interpretations in terms of distribution-derivatives of discontinuous point-functions.

Another example, of great importance in the theory of Harmonic functions, is the so-called fundamental solution of the Laplace equation: $f(x) = \frac{1}{r^{n-2}}$ in case $n \geqslant 3$, and $f(x) = \log \frac{1}{r}$ if $n = 2$. Here r denotes distance from the origin to the point x:

$$r = \{\Sigma_{i=1}^n (x_i)^2\}^{\frac{1}{2}}.$$

This function $f(x)$ is harmonic, that is, its ordinary Laplacian is zero, at each point x except at the origin where the "ordinary" Laplacian does not exist as a point-function. However, this $f(x)$ can be considered as a distribution and its Laplacian distribution easily calculated. We find that this Laplacian distribution is ***not*** the identically zero distribution; it turns out to be $-N\delta_0$ where δ_0 is the n-dimensional Dirac delta-distribution corresponding to unit mass concentrated at the origin; N is a constant which depends on the dimension n:

$$N = \frac{(n-2)2\pi^{n/2}}{\Gamma\left(\frac{n}{2}\right)}$$

if $n \geqslant 3$, in particular, $N = 4\pi$ if $n = 3$, and $N = 2\pi$ if $n = 2$. It is highly significant that this fundamental or elementary solution is really a solution of the non-homogeneous (distribution!) Laplace equation with second member a delta-distribution.

We choose our final example from the theory of holomorphic functions. We consider the case $n = 2, f = f(x, y)$ a complex-valued function and set

$$x = \tfrac{1}{2}(z + \bar{z}), \qquad y = \frac{1}{2i}(z - \bar{z}).$$

z, $\bar{z}$ cannot be considered as independent variables but we define $\frac{\partial}{\partial z}, \frac{\partial}{\partial \bar{z}}$ as follows:

$$\frac{\partial}{\partial z} = \tfrac{1}{2}\left(\frac{\partial}{\partial x} - i\frac{\partial}{\partial y}\right),$$

$$\frac{\partial}{\partial \bar{z}} = \tfrac{1}{2}\left(\frac{\partial}{\partial x} + i\frac{\partial}{\partial y}\right).$$

The necessary and sufficient conditions that $f(x, y)$ be holomorphic in the neighbourhood of a point (x, y) are that f be continuously differentiable and satisfy the Cauchy-Riemann equations in this neighbourhood: $\frac{\partial f}{\partial \bar{z}} = 0$. At a single point, of course, these conditions are meaningless. However, if f determines a distribution F, then the distribution $\frac{\partial F}{\partial \bar{z}}$ can be formed and it will not be the identically zero distribution; this $\frac{\partial F}{\partial \bar{z}}$ can be a useful tool in studying the singularity.

Thus if $f = \frac{1}{z}$ it turns out that $\frac{\partial F}{\partial \bar{z}} = \pi\delta_0$. This distribution theorem could be used in the development of the theory of holomorphic functions, Cauchy integrals and so on. It could also be useful in the more difficult study of functions of several complex variables.

The function $f = \frac{1}{z^m}$ is not summable for $m > 1$, but f determines a distribution through the use of Hadamard "finite part" which coincides in this case with Cauchy "principal value."

$$F = \text{Fp}.\left(\frac{1}{z^m}\right) = \text{pv.}\left(\frac{1}{z^m}\right)$$

and

$$\frac{\partial F}{\partial \bar{z}} = \frac{\partial}{\partial \bar{z}}\left(\frac{\partial}{\partial z}\right)^{m-1}\left\{\frac{(-1)^{m-1}}{(m-1)!}\frac{1}{z}\right\} = \frac{(-1)^{m-1}\pi}{(m-1)!}\left(\frac{\partial}{\partial z}\right)^{m-1}\delta_0$$

(a multiple layer located at the origin).

A general meromorphic function $f(z)$ determines a distribution F if the Cauchy principal value is used at poles of order greater than one. Then $\frac{\partial F}{\partial \bar{z}}$ is a sum of distributions each of which has a supporting set consisting of a single point: the "single points" are the poles of the meromorphic function and the distribution $\frac{\partial F}{\partial \bar{z}}$ expresses the ordinary residue formula in the theory of meromorphic functions.

§10. DISTRIBUTIONS IN SEVERAL VARIABLES; THEORY

FIRST, the expansion lemma of §3 has extensions to n variables, of which the simplest is as follows:

Lemma 1. *Let R be a closed n-dimensional rectangle $a_i \leqslant x_i \leqslant b_i$, $i = 1, \ldots, n$ and let R_0 be the $n-1$ dimensional rectangle $a_i \leqslant x_i \leqslant b_i$, $i = 2, \ldots, n$. Let $\theta(x_1)$ be in $S(a_1, b_1)$ with*

$$\int_{a_1}^{b_1} \theta(x_1)dx_1 = 1$$

and for each ϕ in S_R let

$$\phi_0(x_2, \ldots, x_n) = \int_{a_1}^{b_1} \phi(x_1, \ldots, x_n)dx_1.$$

Then ϕ_0 is in S_{R_0} and

$$\phi = \phi_0\theta + \frac{\partial}{\partial x_1}\rho$$

with a uniquely defined ρ in S_R.

Hence for every distribution F on R there are x_1-primitives G, that is, with $\frac{\partial G}{\partial x_1} = F$, and they can be expressed by the formula

$$G(\phi) = G_0(\phi_0) - F(\rho)$$

where G_0 is an arbitrary (but fixed) distribution on R_0 (thus G_0 is a distribution in the variables $x_2, \ldots, x_n$).

If Ω is an *open* rectangle $\alpha_i < x_i < \beta_i$, $i = 1, \ldots, n$ and Ω_0 is the open rectangle $\alpha_i < x_i < \beta_i$, $i = 2, \ldots, n$ put

$$\phi_0 = \int_{\alpha_1}^{\beta_1} \phi(x_1, \ldots, x_n)dx_1$$

and choose $\theta(x_1)$ in some $S(a_1, b_1)$ with $\alpha_1 < a_1 < b_1 < \beta_1$ and

$$\int_{a_1}^{b_1} \theta(x_1)dx_1 = 1.$$

Then we may again write

$$\phi = \phi_0\theta + \frac{\partial\rho}{\partial x_1}$$

and the x_1-primitives of F on Ω can be expressed by the formula

$$G(\phi) = G_0(\phi_0) - F(\rho)$$

where G_0 is an arbitrary distribution on Ω_0.

Next, with the methods of §5 we show that if F is a c.l.f. on a finite closed rectangle R then $F = f^{(r)}$ for some suitable summable $f(x)$ and some $r = (r_1, \ldots, r_n)$. Exactly as in §5 we begin by proving that there must be a $p = (p_1, \ldots, p_n)$ which depends on F, such that

(C_p) $F(\phi_m)$ converges to $F(\phi)$ whenever $\phi_m{}^{(q)}$ converges uniformly to $\phi^{(q)}$ for each $q \leqslant p$.

Then, since for $q \leqslant p$,

$$\phi^{(q)}(x) = \int_{a_1}^{x_1} \cdots \int_{a_n}^{x_n} \frac{(x_1 - t_1)^{p_1-q_1-1}}{(p_1 - q_1 - 1)!} \cdots \frac{(x_n - t_n)^{p_n-q_n-1}}{(p_n - q_n - 1)!} \phi^{(p)}(t)dt_1 \ldots dt_n$$

it follows that $|\phi^{(q)}| \leqslant K|\phi^{(p)}|$ for all $q \leqslant p$ for some finite K which depends on p and R but not on ϕ. Hence for this p,

(B_p) $F(\phi_m)$ converges to $F(\phi)$ whenever $\phi_m{}^{(p)}$ converges uniformly to $\phi^{(p)}$.

This, in turn, implies, as in §5,

(A_p) $|F(\phi)| \leqslant |F|_p|\phi^{(p)}|$

for all ϕ, for some finite constant $|F|_p$ (we let $|F|_p$ denote the least possible such constant).

Finally, as in §5, we form the linear functional L,

$$L(\phi^{(p)}) = F(\phi)$$

defined for functions of the form $\phi^{(p)}$ with norm $|\phi^{(p)}|$ and extend this functional to all continuous functions by the Hahn-Banach procedure. The Riesz representation theorem then applies: there is a function $\psi(x)$ of total (n-dimensional) variation equal to $|F|_p$ such that

$$F(\phi) = L(\phi^{(p)}) = \int_R \phi^{(p)}(x)d\psi(x)$$

and we may also suppose that $|\psi(x)| \leqslant |F|_p$ for all x. This implies, by an n-dimensional integration by parts,

$$F(\phi) = (-1)^n \int_R \psi(x)\phi^{(p_1+1, \ldots, p_n+1)}(x)dx,$$

so that $F = f^{(r)}$ with $r = (p_1+1, \ldots, p_n+1)$ and $f(x) = (-1)^{p_1+\ldots+p_n}\psi(x)$. (We note that $|f| \leqslant |F|_p$.) If $f(x)$ is replaced by its integral

$$\int_{a_1}^{x_1} \ldots \int_{a_n}^{x_n} f(t_1, \ldots, t_n)dt_1 \ldots dt_n$$

and r is taken to be $(p_1+2, \ldots, p_n+2)$ we obtain $F = f^{(r)}$ with f continuous.

Next we generalize the results of §6. The method used in §6 serves to prove that if F_a are bounded on R then for all a and ϕ,

$$|F_a(\phi)| \leqslant K|\phi^{(p)}|$$

for some p and some finite K independent of a, ϕ, that is, the $|F_a|_p$ are bounded. As in §6 it follows that the F_a on R are bounded if and only if they can be expressed in the form

$$F_a = (f_a)^{(r)}$$

with $|f_a|$ bounded, or equivalently (with different f_a and r) with $f_a(x)$ a set of equicontinuous point-functions. From this we conclude that if F_m is convergent, its limit $F(\phi) = \lim_{m\to\infty} F_m(\phi)$ must be a c.l.f.; we can conclude also that there must be a representation $F_m = f_m^{(r)}$ with $f_m(x)$ a uniformly convergent sequence of continuous functions from the following lemma.

Lemma 2. *If $g_m(x)$ are equicontinuous and for some fixed r and every ϕ,*

$$\int_R g_m(x)\phi^{(r)}(x)dx$$

converges to zero as m becomes infinite, then there are equicontinuous functions $P_m(x)$ of the form

$$P_m(x) = \Sigma_{i=1}^{n} \Sigma_{j=0}^{r_i-1} h(x_1, \ldots, x_{i-1}, x_{i+1}, \ldots, x_n)x_i^j$$

such that the $g_m(x) + P_m(x)$ converge uniformly to zero.

To prove this lemma consider the case $r = (0, \ldots, 0)$ exactly as in §6 but with intervals replaced by n-dimensional rectangles and then complete the proof by use of induction on the indices, one at a time.

As for the partition theorem of §7, it can be proved for finite closed n-dimensional rectangles by using products of functions $a(x_i)$, $i = 1, \ldots, n$, with each a of the type described in §7. It then follows that the local character determines the distribution uniquely and, in particular, the zero-distribution is the only one with nullity-set which includes all x.

§11. The convolution product

We now define two bilinear products. First, the direct product of two distributions; for locally summable point-functions $f(x)$, $g(y)$ the direct product of of $f(x) \times g(y)$ should reduce to the usual product $f(x)g(y)$ which is a locally summable function in two variables. This suggests that for distributions S_x, T_y over the spaces of x, y respectively, $S_x \times T_y$ should be defined so that $S_x \times T_y \,.\, \phi(x, y) = S_x u(x) T_y v(y)$ whenever $\phi(x, y)$ is the product of testing functions u, v in x, y respectively. Now it can be shown that this requirement leads to a unique $S_x \times T_y$. Moreover, for a general testing function $\phi(x, y)$ it can be shown that $T_y \,.\, \phi(x, y)$ is a testing function in x and that

$$S_x \times T_y \,.\, \phi(x,y) = S_x \,.\, (T_y \,.\, \phi(x,y)) = T_y.(S_x \,.\, \phi(x,y)).$$

For example, if δ_x, δ_y are Dirac δ's in x space, y space respectively then $\delta_x \times \delta_y$ is a δ in (x, y) space.

We shall use the direct product to define the more important convolution product. For point-functions the convolution product $h = f{*}g$ is well known and

$$h(x) = \int f(x - t)g(t)dt = \int f(t)g(x - t)dt.$$

However local summability of f and g is not sufficient to ensure that h exists and is locally summable. This difficulty disappears if at least one of f, g has a bounded supporting-set.

For example if $B(x)$ is 1 for $|x| < 1$ and 0 elsewhere, then at each x the value of $f * B$ is the average of f over all t with $|x - t| < 1$.

To extend the definition of convolution product to distributions we first write

$$\begin{aligned} h(\phi) &= f{*}g \,.\, \phi = \iint f(x - t)g(t)\phi(x)dxdt \\ &= \iint f(\xi)g(\eta)\phi(\xi + \eta)d\xi d\eta \\ &= f_\xi \times g_\eta \,.\, \phi(\xi + \eta). \end{aligned}$$

This suggests for distributions

$$(S * T)_x \,.\, \phi(x) = S_\xi \times T_\eta \,.\, \phi(\xi + \eta).$$

If at least one of S, T has a bounded supporting-set it can be shown that this formula leads to a unique $S * T$ even though $\phi(\xi + \eta)$ does not vanish outside a finite rectangle in ξ, η space (that is, $\phi(\xi + \eta)$ is not a testing function).

This definition leads to identities: $\delta * T = T$, $\delta' * T = T'$, $D^p T = D^p * T$. We use D^p to represent both the derivation operator and also the distribution $D^p\delta$; we use the Laplacian Δ in the same way. In general, $(S * T)' = S' * T = S * T'$. It follows from this that if α is a testing function then $T * \alpha$ (the regularization of T by α) can be formed, since α may be considered as a distribution with bounded supporting-set, and this $T * \alpha$ is always a point-function with point-function derivatives of all orders. This leads to the theorem that every distribution is a limit of testing functions (considered as distributions).

In a space of n dimensions where $n > 2$, it is customary to define the potential U^f associated with the mass-density $f(x)$ as follows:

$$U^f = \int \frac{f(t)}{|x - t|^{n-2}} \, dt$$

($|x - t|$ = Euclidean distance $r = \sqrt{\Sigma(x_i - t_i)^2}$). This extends to distributions, in the form

$$U^T = T * \frac{1}{r^{n-2}}$$

and gives the formula of Poisson

$$\Delta U^T = U^{\Delta T} = \Delta * \frac{1}{r^{n-2}} * T = -N\delta * T = -NT$$

where N has the same value as in §9.

The use of convolutions with distributions has important applications in the problem of Cauchy in partial differential equations.

§12. Fourier series for distributions

For simplicity we shall take the period 1 instead of 2π and let x correspond to the points of a circle so that the points $x = 0$ and $x = 1$ are identified. A testing function is now a continuous function $\phi(x)$ with continuous derivatives of all orders such that ϕ and its derivatives have the same values at $x = 0$ as they do at $x = 1$. We shall say that $\phi_m \to \phi$ if for each p, $\phi_m^{(p)}(x)$ converges uniformly to $\phi^{(p)}(x)$. A distribution T is now a linear functional defined for all such ϕ and such that $T . \phi_m \to T . \phi$ if $\phi_m \to \phi$ in this sense. In the present situation every distribution T is identifiable with a finite sum of continuous point-functions and derivatives of such continuous point-functions. The usual calculus of distributions applies but now convolution $S * T$ is defined for all S, T.

First, we define Fourier coefficients as follows. For a function $f(x)$ we have

$$a_k(f) = \int_0^1 f(x) \, e^{-2i\pi kx} \, dx.$$

Generalizing this formula for distributions, we define

$$a_k(T) = T . e^{-2i\pi kx}.$$

Then every distribution T has a sequence of Fourier coefficients.

Secondly, for each distribution there is a polynomial in k, $P(k)$, such that $|a_k(T)| \leqslant P(k)$. Indeed, if T is a continuous function f, its Fourier coefficients are bounded and if $T = f^{(p)}$ then

$$a_k(T) = (2i\pi k)^p a_k(f).$$

Thirdly, the Fourier series $\Sigma a_k(T)e^{2i\pi kx}$ is always convergent and converges to the distribution T. Let us prove this first for the special case of the δ distribution. Here the coefficients $\delta(e^{-2i\pi kx}) = e^0 = 1$ and we need to show that $\Sigma e^{2i\pi kx} = \delta$. But this is so since for every testing function ϕ,

$$\Sigma e^{2i\pi kx} \,.\, \phi = \Sigma a_{-k}(\phi) = \phi(0) = \delta(\phi).$$

Now for a general T we have

$$\begin{aligned} \Sigma a_k(T)e^{2i\pi kx} &= \Sigma(T_\xi \,.\, e^{-2i\pi k\xi})e^{2i\pi kx} \\ &= \Sigma T_\xi \,.\, e^{2i\pi k(x-\xi)} = \Sigma T * e^{2i\pi kx} \\ &= T * \Sigma e^{2i\pi kx} = T * \delta = T. \end{aligned}$$

It can be shown that every trigonometric series with coefficients bounded by a fixed polynomial converges to some distribution with the given coefficients as its Fourier coefficients.

INDEX

www.ingramcontent.com/pod-product-compliance
Ingram Content Group UK Ltd.
Pitfield, Milton Keynes, MK11 3LW, UK
UKHW032122310125
454513UK00004B/154

9 781487 591328